William Samuel Waithman Ruschenberger

A Report on the Origin and Therapeutic Properties of

Cundurango

William Samuel Waithman Ruschenberger

A Report on the Origin and Therapeutic Properties of Cundurango

ISBN/EAN: 9783337811174

Printed in Europe, USA, Canada, Australia, Japan

Cover: Foto ©berggeist007 / pixelio.de

More available books at **www.hansebooks.com**

A REPORT

ORIGIN AND THERAPEUTIC PROPERTIES

OF

CUNDURANGO.

BY

W. S. W. RUSCHENBERGER, M. D.,

MEDICAL DIRECTOR U. S. NAVY; PRESIDENT OF THE ACADEMY OF NATURAL
SCIENCES OF PHILADELPHIA, ETC., ETC.

Published by order of the Navy Department.

GOVERNMENT PRINTING OFFICE.
1873.

United States Naval Hospital,

Philadelphia, December 23, 1872.

Sir: In April last I had the pleasure to present to the Surgeon-General of the Navy a report on cundurango, the materials for which had been submitted to me some time previously by the Chief of the Bureau of Medicine and Surgery of the Navy Department.

A large part of the information relating to the history of cundurango which is contained in that report has been derived from Dr. Ayers, whose herbarium and report are herewith returned. In the botanical study of the specimens I am greatly indebted to Dr. Joseph Carson, Professor of Materia Medica in the University of Pennsylvania, who is distinguished for his knowledge of medical botany.

Although experience has shown that a specific remedy for cancer has not been found in cundurango, the merit of collecting, through the labors of Passed Assistant Surgeon Ayers, the information contained in the report, is not abated. The subject has not yet lost its interest. Had the promised efficacy of cundurango been fully realized, the prompt effort of the Bureau to obtain accurate knowledge of the plant, and make it common to the public, would have been generally commended.

Under an impression that the medical profession would be gratified to know what the Bureau of Medicine has done in the premises, I beg leave to invite your attention to the report.

I am, very respectfully, your obedient servant,

W. S. W. RUSCHENBERGER,
Medical Director United States Navy.

Surgeon-General
JAMES C. PALMER, U. S. Navy,
Chief of Bureau of Medicine and Surgery,
Navy Department.

CUNDURANGO.

The attention of the medical profession and people of the United States was first attracted to cundurango as an efficient remedy for cancer through the State Department of the Government of the United States.

It is stated that Dr. Casares, of Quito, in September, 1869, administered cundurango in a case of cancer. The result of the treatment in a few days was so favorable that he brought it to the notice of his government, although he was unable to continue the use of it at that time because the small quantity of it in his possession had been consumed. Subsequently it was procured in larger quantity.

Relying upon the accuracy of Dr. Casares, the government of Ecuador determined to send fifty pounds of cundurango to scientific societies of Paris and London for analysis, and also a circumstantial statement of its therapeutic effects, which Dr. Casares was requested to make. In answer to this request, he reported April 28, 1870, to the Minister of the Interior of the government of Ecuador, several cases of cancer cured through the agency of this plant. The descriptions of these cases, published in No. 425 of El Nacional—the official journal of Ecuador—are somewhat obscure. In the translations issued here, they appear as "Extracts from the reports of Doctors Casares, Eguiguren," and others. In illustration of their obscurity, the following are quoted:

"Passing to another case, I will state that it is a year since José A—— entered this hospital, suffering from intermittent fever and scrofulous ulcers in the breast. Constant and diligent care did nothing toward restoring his health until I resolved to administer the cundurango to him together with arsenic, and his present condition is very encouraging.

*   *   *   *   *   *   *   *

" I will remark that the cancers which I treated with the cundurango were already in the form of fungus hæmatodes and none in the state of rawness.

*   *   *   *   *   *   *   *

"Reminding the patient of this terrible circumstance, (the difficulty of saving his life,) I obliged him to take the remedy which I alone possessed in Quito. After a few days it produced so surprising an effect that Dr. Casares was astonished at the rapidity with which the cure took place, until the edges became cicatrized and also the bone, as if it had been a wound in soft parts and in a healthy subject. I caused Dr. Casares to be informed of the remedy which had cured him, and this professor appreciated a medicament of such high importance ; and afterward, learning that a servant of my brother Manuel Eguiguren had cured himself some months before of an ulcerated cancer which resisted the known appliances of art, he began to give it to a patient who was by chance in the hospital, and who would have died two or three days afterward, if this remedy had not been so seasonably given him, as I was assured by Dr. Casares, when he took me to see him."

The above seems to be a statement of Dr. Eguiguren. He asserts that he only possessed the remedy, and that he "caused Dr. Casares to be informed" of its efficacy. The remark that none of the cancers which he treated was "in a state of rawness" but "already in the form of fungus hæmatodes" is not consistent with the views of the progress of the disease commonly entertained by surgeons

The Hon. Rumsey Wing, the American minister near the

government of Ecuador, addressed Mr. Fish, Secretary of State, letters on the subject, dated Quito, January 5, and 25, 1871, in which he states, substantially, that for some time past he had been aware of "the *recent* discovery" of a specific remedy for cancer and other formidable diseases. He says that the discovery is ascribable to "one of those *trivial incidents* which evince oftentimes how humble are the means used by Providence to make known the mysteries of nature and the wonders of science." He states that "the ignorant wife of a common Indian laborer, who had for some time suffered most fearfully from an internal cancer, concluded finally," "in perfect good faith and honesty of purpose, and with no sinister or malevolent design" "to administer eternal relief to him by the simple process of mixing poison in his food!"

The agent she selected for the purpose of releasing her husband from his agony was the fruit of the cundurango, "said to be an active poison;" but not being able to obtain it, she administered instead a decoction of the wood. To her surprise, the man at once began to rally. She continued to dose him in augmented quantities, and he entirely recovered. The case becoming public, "further tests have established the powers of the wood as a remedial agent."

The Hon. A. Flores, the minister from Ecuador to the United States, informed the Secretary of State, in a letter, dated Washington, D. C., March 14, 1871. that his government had forwarded for his acceptance a box of cundurango, "to which great medicinal qualities are attributed," as he would see in "the annexed extracts copied from No. 425 of the official journal of Ecuador." Mr. Flores desired Mr. Fish, in the event of his receiving it, to enable him to communicate to the government of Ecuador "the result of the analyses and experiments" which might be made.

The letters referred to above were made public. The cundurango was distributed through the Smithsonian Institution, and the medical departments of the Army and Navy, with a statement of a method of using it, communicated by the Hon. Rumsey Wing to the Secretary of State, which is as follows:

"A decoction of cundurango is made out of a small piece of the wood beaten out flat, and a half an ounce of which is boiled in three tea-cups of water.

"Of this decoction one tea-cupful must be taken in the morning and another at night.

"If the patient has ulcers, they should be kept very clean by the use of aromatic wine or simple ointment, according to the degree of inflammation.

"The decoction is only to be used for fifteen days consecutively. An interval of fifteen days must then elapse, when the remedy may again be resumed for the same length of time.

"Great attention must be given in order to keep the digestive organs in the best possible condition.

"With regard to any other counsel, the general aspect of the patient must serve as guide.

"In this country, [Ecuador,] the cundurango has proved itself to be a powerful restorative, and must eventually work a complete revolution in the treatment of cancerous, venereal, and ulcerous diseases.

"DR. CASARES.

"QUITO, *January* 4, 1870.

"A true translation:

"RUMSEY WING."

Influenced by the high authority under which it was introduced to public notice, Surgeon-General William M. Wood, of the Navy, immediately, May 15, 1871, instructed Passed Assistant Surgeon Joseph G. Ayers, United States Navy, then attached to the United States Steamer Resaca, on the coast of

Peru, to collect botanical specimens of cundurango and such
information relating to it as might be possible for him to
obtain.

The Resaca arrived at Guayaquil (latitude 2° 30′ south)
August 10, 1871. There Dr. Ayers was unable to obtain any
cundurango, or much reliable information concerning it. He determined to visit the province of Loja, where the plant grows, and
started August 13th, with three companions, one in the capacity
of interpreter. On the 15th he landed at the little town of
Santa Rosa, about ninety miles south of Guayaquil, whence he
proceeded up the valley, over very difficult roads, toward
Zaruma. He crossed a mountain-range 10,000 feet above the
level of the sea, and in the morning of the 19th reached the
scattered hamlet of Ayapamba, in the deep valley of the Calera
River, where, for the first time, he saw the cundurango growing.
The next day he arrived at Zaruma, which is 4,000 feet above
the sea and about fifty-four miles from Santa Rosa. This is
regarded as the center of the cundurango district.

He found here a great variety of vines, including the cancer
remedy, of which little had been sent to the United States,
though considerable quantities of other vines named cundurango, and among them one called Bejúco pachón, had been
shipped.

On the 24th of August he left Zaruma, crossed the mount
ains in a southeasterly direction, and the next day reached the
small village of Cisne, whose inhabitants he found busy
gathering cundurango from the neighboring valleys for the
Loja market.

The following morning he descended to the extensive plain
of La Toma, through which the river Catamayo flows toward
the Pacific. This plain is many miles in extent and is inclosed
by high mountains, which at this season are bare or thinly
covered with dry and faded grass. The appearance of this

region is exceedingly sterile. Except in the deep and narrow valleys between the mountains, and along the banks of the river, there is but little verdure. Cundurango grows in this locality.

Leaving the plain, he crossed a mountain-range at an elevation of 9,700 feet, and through the beautiful and fertile valley of Loja, to the town of that name near to its eastern side. This valley, surrounded by mountains, is twelve miles long and seven wide.

The town of Loja, (latitude 4° south and longitude 79° 24' west), a place of considerable wealth and refinement, is 7,400 feet above the level of the sea, and about one hundred and twenty-five miles from Santa Rosa. Its population, chiefly of Spanish and aboriginal descent, is estimated at 4,000.

A Dr. Keene had been here from the United States to purchase cundurango, a little of which grows within ten or fifteen miles of the town, but is found in the valleys of less elevation and more tropical temperature in the vicinity. People were cutting it in all the neighboring country. Several thousand quintals had been brought already to the city, and it was estimated that there are 40,000 quintals in this part of the province. A quintal is a hundred pounds.

August 31st he descended the valley of El Rio de los Andes to the village of Malacatos, twenty-one miles farther south. Nearly all the genuine cundurango sent to the United States is from this place. Here, as well as elsewhere, it was difficult to find a large plant still standing, so thoroughly had it been gathered for the market. At the present rate of cutting it, there will be in a short time but little left in the province.

Having visited Vilcabamba, La Toma, and Chapamarca, he returned by the way of Zaruma, and rejoined the ship September 21st. Dr. Ayers, in this journey of thirty-nine days, collected specimens of cundurango leaves, flowers, and fruits of several plants, and much information, which is embodied in his report.

The collections forwarded by him to the Bureau of Medicine and Surgery contain specimens labeled—

Cundurango blanco.
Two varieties of cundurango de palóma.
Cundurango de tumbo grande.
Cundurango de plátano.
Cundurango de tumbo chico.
Cundurango cascarilla.
Cundurango saragosa. *(Sic.)*
Cundurango amarillo.
A variety of cundurango growing near Guayaquil.

Of these ten plants, called cundurango, he procured flowers of those only which bloom in August and September: most of them flower in the month of May.

The word cundurango is probably derived from the Quíchua, the language of the Inca races. According to a vocabulary of the language before me, the word *cuntur* signifies vulture, and *anca*, eagle. *Cuntur* has become *condor* in the mouths of the Spanish conquerors, and designates a particular vulture. Whether the term is a compound of *cuntur* and *anca* or not, and signifies condor vine, as has been stated, it is worthy of note that some ten plants are designated in a generic sense by a Quíchuan name, implying that they had probably attracted attention of the aborigines of the region before the conquest.

CUNDURANGO BLANCO—white cundurango—is the plant which has recently acquired so much notoriety as a remedy for cancer. Careful examination of the specimens furnished lead to an opinion that it has not yet been described. The materials at hand are not sufficient to enable us to characterize it botanically and give it a systematic name.

This view is sustained in a letter, dated at Guayaquil, December 13, 1871, and addressed to the editor of "Nature," by Dr. A. Destruge. He says: "The cundurango belongs to the order Asclepiadaceæ, third tribe, which corresponds to Asclepiadeæ veræ, first division, Astephanæ, whose characters are that the limb of the corolla is without scales, and the stamens without appendage or corona.

"This division comprehends only five genera, viz: Mitostigma, Astephanus, Hæmax, Hemipogon, and Nautonia. In none of these genera can the cundurango be classed.

"The genus Mitostigma as a distinguishing character has two long filaments at the end of the stigma, and this is not the case in cundurango. The genus Astephanus has the sepals acute, the carolla subcampannlate, and the stigma elongated; characters that do not belong to the cundurango. The genus Hæmax has the divisions of the corolla hooded, and other characters not observed in the cundurango. The genus Hemipogon has the sepals of the calyx acute, hard, with a curved extremity. The corolla is campannlate, which is not the case in cundurango. The genus Nautonia has the sepals striated and curved, which is not the case in cundurango.*

"The flowers of the cundurango have a calyx of five divisions, obtuse, ovate, and villose in their inferior part and of quincuncal pre-efflorescence. The corolla is rotate, of five divisions, lanceolate, hairy at the base on the inside, and somewhat fleshy, with a membraneous margin. Its æstivation is imbricated. The stamen has no appendage or corona; the anthers are terminated by a membrane; and the pollen-masses are elongated and suspended. The stigma is pentagonal and conical. The flowers are numerous and disposed in umbelliferous inflorescence.

"As aforesaid, the cundurango forms a new genus."

The fruit is a follicle, about four inches long and an inch in

* De Candolle, Prodr. pars viii, p. 507.

diameter, deeply ribbed longitudinally, and, in the dried condition, ash-colored. It contains numerous black seeds armed with long silky aigrettes imbricated on the placenta.

Dr. Destruge has not given a name to this genus. I suggest that it be called Pseusmagennctus equatoriensis, (from ψευσμα, a lie, a fraud, and γεννητης, a parent, a producer.)

The mature vine of the cundurango blanco is from ten to thirty feet in length, less than one inch in diameter, often more, but rarely exceeds two inches. The external surface of the bark is generally smooth, of an ash-gray color, more or less mottled with greenish or blackish lichens. On the lower part of the stem it is frequently reddish-brown. These characters are retained in the dried bark. The wood is white, with a faint, yellowish tint, and possesses numerous minute round pores and a slender pith. The bark and pith contain much, but the wood very little, milky juice.

The bark is prepared for market by pounding the stem with a mallet to detach it, and then drying it in the sun, generally on skins, during eight or ten days. It assumes the form of quills, or semi-cylindrical pieces, of from one to four inches long, often very much broken from pounding it into the sack to reduce the bulk of the package. The bark is from one-sixteenth to one-sixth of an inch thick. The liber is yellowish-white, and, in transverse section, presents numerous minute yellowish points, which, in longitudinal section, are considerably elongated. The bark dried upon the stick is of a much darker color than that which is dried after it is detached. Its taste is bitter and aromatic, and, when dried, without odor.

The milky juice, which all parts of the plant contain, when evaporated to dryness, has a resinous appearance.

This plant grows in localities from 3,000 to 5,000 feet above the level of the sea, but generally at an elevation of about 4,000 feet.

A chemical analysis of the wood and bark by Dr. Antisell has been published in the Medical Record, New York, December 15, 1871. It is as follows:

Ratio of wood and bark.

"Bark... 49.72
"Wood... 50.28

"Average of three examinations................... 100.00

"Constitution of bark.

"Moisture...................................... 8 parts
"Mineral salts................................ 12 "
"Vegetable matters........................... 80 "

100 "

"These vegetable matters were separable by the usual methods into the following:

"Fatty matter, soluble in ether and partially in strong alcohol.. 0.7
"Yellow resin, soluble in alcohol.................... 2.7
"Starch, gum, and glucose......................... 0.5
"Tannin, yellow and brown coloring matter, and extractive... 12.6
"Cellulose, lignin, &c................................ 64.5

81.0

"On distillation, no volatile oil or acid was obtainable; no crystalline alkaloid or active principle was separable by the usual method of proximate analysis. Whatever medicinal virtues the plant may possess must reside either in the yellow resin or in the extractive. The former is soluble in alcohol, the latter in water. In the water decoction some of the resin is diffused; but the greater portion of resin is not extracted by water. The therapeutic position of the plant, judged from analysis, is among the aromatic bitters.

The CUNDURANGO DE TUMBO GRANDE, (big-fruit cundurango,) also called Bejúco pachón,* grows from twenty to fifty feet high, and is about an inch in diameter near the ground. Externally the bark on the lower part of the stem is of dark gray color and fissured both longitudinally and transversely. Toward the top it is smooth, of an ash-gray color, and closely resembles that of the cundurango blanco. The liber is yellow or yellowish white. The wood, which is pale yellow or white, has numerous round pores and a small pith. The bark dried for the market is from one-sixteenth to one-sixth of an inch thick, in the form of quills, or semi-cylindrical, more or less irregular, pieces of from one-eighth to three-fourths of an inch in diameter. The greater part of the outer and dead layers are removed in detaching it from the stem with a mallet. The liber becomes dark brown or nearly black in drying, and there are tubercules a line or more in diameter on its external surface. The taste of the bark is moderately bitter and aromatic.

Except the wood, in which there is little, all parts of the plant contain a large quantity of milky juice.

The fruit is a follicle from six to eleven inches long, and from about three to seven in diameter. Dr. Ayers says it closely resembles a water-melon. It contains a large number of black seeds armed with long silky aigrettes which are imbricated on the placenta.

---

* Bejúco pachón. Bejúco is a name applied in Spanish America to twining-plants which may be used like osiers, or as wythes. Don Jorge Juan and Don Antonio de Ulloa, in their Noticias Secretas de America. mention, page 571, a "bejúco," known as the Carthagena bean, which is esteemed as an efficacious antidote against the bite of every kind of viper and poisonous animal.

Pachón signifies hunting-dog, canis venaticus. The word is used also to designate a man of morose temper and dull intellect; a tardy person. The significance of the term in connection with this cundurango is not apparent.

This plant grows in considerable abundance near Zaruma, and sparingly in the vicinity of Loja.

CUNDURANGO DE TUMBO CHICO (small-fruit cundurango) very much resembles the cundurango de tumbo grande in length, size, wood, and external appearance. Its bark is deeply grooved longitudinally, sharp ridges standing out from one-eighth to one-half inch betwixt the grooves. It is very light and friable. On the lower part of the stem its color is dark chestnut-brown or blackish; but near the top it is gray and but little fissured. The liber is white. The wood is pale yellow or white. All parts of the plant, except the wood, contain a large quantity of milky juice. It is difficult to distinguish the dried barks of the two plants (grande and chico) from each other. The liber of the tumbo chico is lighter in color; often nearly white. It has no tubercules, and to the touch the outside is smooth and soft, owing to the friable corky layer. In the tumbo grande this layer is hard-feeling laminæ broken by transvere cracks. It has a sweetish, pleasant taste.

The follicle of the tumbo chico is from six to eight inches in length and about an inch and a half to two inches in diameter.

It grows in the same localities as the tumbo grande.

CUNDURANGO DE PALÓMA, (pigeon cundurango.)—There are two plants of this name. One grows near Malacatos and Loja, and the other in the vicinity of Zaruma. Both are small, generally less than one-half inch in diameter. Both have a porous wood and a white liber, the taste of which is sweet. Both plants, except in the wood, contain an abundance of milky juice. They grow at the same elevations as the other cundurangos.

CUNDURANGO DE PALÓMA from Zaruma.—Echites hirsuta: Ruiz and Pavon., Flor. Peruv., vol. 2, page 19, tab. 136. Echites pavonii: De Candolle, Prodromus, tome 8, page 463. This plant is fruticose, twining, and hirsute. The stem is branching, contorted, and striated. The branches are uniform. The leaves are from four to six inches in length, oblong oval, acuminate, densely villous below, excavated at base with one or more oblong glands. The petioles are round, striated, and curved. The racemes are axillary, solitary, and simple, longer than the leaves, pedicles one-flowered, tracts lanceolate-acuminate, deciduate-stipulate. Calyx in flower acute; in fruit, obtuse. Color, yellow. Flowers in August and September. Habitat in Peru.

The bark has a corky layer of pale yellow or white color, a line or more in thickness. It is longitudinally fissured, showing the dark-green cellular envelope beneath.

CUNDURANGO DE PALÓMA from Malacatos.—Gonólobus tetragonus: De Candolle, Prodromus, tome 8, page 594. Cynanchum tetragonum: Flor. Flamin., tab. 69. This is a vine-like plant with pubescent branches; the leaves are ovate-lanceolate, subcordate, acuminate, and somewhat acute. Peduncles equaling the middle of the petiole; two-flowered; pedicels elongated, shorter than the leaf. Sepals ovate, somewhat acute; petals ovate, obtuse, smooth, and finely venose; follicles ovate-acuminate, four-angled, and indentated, cristate. Seeds, testadenticulate inferiorly. (Decaisne, in Prod. De Candolle, Asclepiadeæ.)

The bark is dark gray or brown, and on the dry stem is slightly wrinkled longitudinally.

CUNDURANGO DE PLÁTANO, plantain cundurango.—Echites acuminata: Ruiz and Pavon., Flor. Peruv., vol. 2, page 19, tab. 134; De Candolle, tome 8, page 449, Apocyneæ.

2 c

A fruticose voluble, glabrous plant, the stem of which is very branched, long, and round. The leaves are three or four inches long, elliptical, obtuse at base, at the apex shortly acuminate, coriaceous; the limb glandular at base; branches axillary, shorter than the leaf; 8-10 flowered; pedicels longer than the calyx; the lobes of the calyx elliptical obtuse: the tube of the corolla quadruple the length of the calyx: the lobes ovate-oblong. Follicles erect, nine inches long, terete. Habitat in Peru. (Adolph De Candolle, in Prodr.)

Cundurango de plátano is a small vine, not often exceeding ten feet in length, or a quarter of an inch in diameter. The outside of the bark is pale gray. The wood is whitish. It is found growing in small quantities near Malacatos and Chapamarca.

The sample of CUNDURANGO CASCARILLA consists of several pieces of stem, from about three-quarters of an inch to two inches in diameter. The bark is comparatively thin, reddish gray in color, without taste, and distinguished from all the others by minute points of a sulphur-colored substance, which is probably a dried exudation. It grows near Zaruma.

CUNDURANGO AMARILLO (yellow cundurango) grows near Zaruma. The bark is reddish brown, thicker than the preceding, comparatively rough, and nearly tasteless.

The variety of cundurango which grows near Guayaquil is very light, of a pale ashy-gray color, and has a slightly bitter taste.

CUNDURANGO SARAGOSA is about three-eighths of an inch in diameter, longitudinally ribbed, of a gray color, and flattened as if it had been compressed while growing. When the stem is freshly broken, it emits a strong, sickening, herbaceous odor, which is very persistent. All the other cundurangos in the dried state are odorless. This one grows in the vicinity of

Guayaquil, and is used as an antidote to the poisonous effects of snake-bites.

The dried bark of these plants is packed in skins, and the wood in a coarse cloth or sacking. Each package contains about one hundred pounds, or a quintal. A mule carries two such packages.

The cundurangos of Zaruma are sent to Guayaquil for exportation by way of Santa Rosa. The journey to the latter place requires three days. Those of the Loja district are transported to Payta. The distance is much greater, but the roads are better. Laden mules are nine or ten days passing over this route.

Of the physiological effects of the cundurango blanco very little has been ascertained. Dr. Ayers ate two or three drachms of the bark in the course of an hour without any sensible effect.

In a case under his observation vomiting and purging occurred after taking two drachms of the bark of the root; whether they were produced by it, he was not satisfied. The reports on its effects upon the lower animals are very contradictory. It was generally said, however, that the root and seeds would kill dogs. For this reason the cundurango blanco has long been called *mata-perro*, dog-killer.

Dr. H. Chiriboga, physician and inspector of the military hospital at Guayaquil, found in his own person that an infusion of four ounces of the stem produced a strong stimulant effect.

A warm infusion of from two to three drachms of the bark in a pint of water was followed by copious diaphoresis and a considerable increase in the activity of the circulation, and, in a short time, by a notable increase of urine, which had the odor of cundurango. In several cases the elimination of urea was augmented; in some, vertigo, disturbance of vision, nervous agitation, and a tendency to delirium were noted.

A patient in the naval hospital at Philadelphia, suffering from syphilis, took, in drachm doses three times a day, sixteen ounces of a fluid extract, prepared at the naval laboratory, New York, from the cundurango sent by Dr. Ayers, without any observable effects of any kind.

Testimony on the therapeutic value of this plant is not satisfactory.

According to his excellency Don Manuel Eguiguren, governor of the province of Loja, cundurango has been used from time immemorial as a special poison for certain animals, dogs and birds, but no medicinal virtues were ascribed to it until Manuela Sanchez, the wife of one Custodio Leon, of the parish of Malacatos, determined to poison him with it, as the readiest means of terminating his sufferings from sores which emitted an intolerable odor and were covered with maggots, as well as to release herself from the duty and fatigue of nursing a hopeless case. He began to improve, from the first dose, and rapidly recovered under its persistent administration.

This is substantially the same tradition which is detailed by the Hon. Rumsey Wing, who states, however, that the patient was afflicted with an "internal cancer," recognized by the Indian wife!

But the Archbishop Rio-frio informed Dr. Ayers that "Custodio Leon was cured of a leprosy which disfigured him."

There is a tradition, very generally received among the people of Malacatos, that he was cured of syphilis of severe character. But whatever the disease might have been, Custodio Leon had been dead, according to the most reliable authority, more than forty years at the time Dr. Ayers visited the parish of Malacatos.

It is asserted in Loja that cancer, rheumatism, dysentery, dysmenorrhœa, chlorosis, and catarrh have been successfully treated with cundurango.

The following testimony is derived from Dr. H. Chiriboga, of Guayaquil:

In a case of " cancer of the right breast, which was ulcerated in the whole extent of the areola," an infusion of cundurango was used twenty-one days. The ulcerated surface was reduced one-half, the pain had disappeared, and the flexibility of the surrounding tissues was increased. From that time the doctor lost sight of this patient.

A man eighty years of age had cancer of the tongue, in an ulcerated condition, for eighteen months. More than half of the organ was destroyed; the tissues at the base of the jaw and in the parotid region were extensively indurated, and in the latter situation ulceration had commenced. He was treated with an infusion of the bark and with local applications of it in form of paste, and was cured in about seven months. But some time afterward he died in a paroxysm of asthma, from which he suffered before he had cancer, while convalescing from an attack of remittent fever.

Mrs. U——, aged 55, had a cancer of the right breast six years, which had been in an ulcerated condition one year. She was treated with an infusion of the wood, without the bark, of cundurango, both internally and locally, during ten months, without benefit. The breast had been nearly destroyed by ulceration ; the glands of the axilla were indurated. Dr. Ayers saw this case before he went to Loja, but not after his return. Dr. Chiriboga informed him, however, that, having used one-ninth of an ounce of the bark in infusion three times a day for fifty days, her general condition had improved, the swelling of the axillary glands had diminished, the pain was less, and although there had been no cicatrization the ulcerated surface seemed more healthy.

Mrs. A—— had a cancer of the left breast, which was removed by surgical operation. Two years afterward the disease returned in the cicatrix and axillary glands, attended with severe pain. It had existed in this condition a year. After she had used cundurango fifty-two days, taking one-third of an ounce of the bark in infusion and eight or ten grains of an alcoholic extract daily, the pain had ceased, the disease had nearly disappeared, and the tumor of the axilla was reduced to one-third of its original size.

Dr. Chiriboga had used cundurango successfully in asthma, and in many cases of rheumatism.

Dr. Edward Moises Costa stated to Dr. Ayers that many cures had been effected with cundurango in the hands of non-medical persons, and that he had used it little more than a year in the hospital of Loja.

A woman between twenty-five and thirty years of age had for a year a pustular eruption, and a large ulcer over the fourth and fifth ribs of the left side, and ulceration of the palate; and for three months a purulent discharge from the vagina, with pains in the thighs and back. On the fifteenth day of treatment the ulcer of the palate and the pains had disappeared, and the vaginal discharge was diminished. The use of the cundurango was suspended. Salines were administered on the fourteenth, fifteenth, and sixteenth days. The cundurango was resumed on the seventeenth day, and continued for seventeen days. On the twelfth day the vaginal discharge ceased, and the ulcer on the ribs was healed. On the forty-fifth day from the commencement of the treatment the patient was discharged cured.

A woman considerably emaciated, who for three months had a fœtid vaginal discharge, was cured in twenty-five days with cundurango, using vaginal injections at the same time.

Two women suffering from blenorrhagia, one during two years, the other during seven months, were cured: one in thirty-seven and the other in one hundred and five days.

A woman, who, on admission into the hospital, had her left arm flexed, without power to extend it, and who moved her lower limbs with difficulty, was discharged cured after three months' treatment with cundurango.

A man who had an ulcer on the left ala of the nose, and another on the anus, of nine months' duration, was cured in thirty days.

A man having an ulcer on the penis was cured in eighteen days.

An extensive ulceration of the right leg of severe character, of two years' duration, was cured in four months.

A woman who had used mercury, and had been unable to cure herself of an ulcer in the throat and severe pains in her head and limbs, was improved in eleven days.

A man, who had been six years unable to work, and for a year unable to mount a horse on account of paralysis, ascribed to sleeping in damp places while in the army, was notably improved after a month's treatment.

A man, forty years old, who had a cancer in the scapular region, which had been in a state of ulceration six months, was cured.

Dr. José M. Eguiguren had used cundurango in many cases of constitutional syphilis in all its stages, and obtained radical cures in more or less time. He used it internally in powder, and in infusion locally. In rheumatic affections as well as in many cases of leucorrhœa, amenorrhœa and ulcers in the vagina, the effects were prodigious. In one case of malignant dysentery, which had resisted all the resources of art, the effect of cundurango was marvelous.

Dr. Augustine Ruiz reports that while in charge of the hospital at Loja in 1863, a negro boy, about thirteen years old, was admitted with a chronic swelling and suppuration of the lymphatic glands of the left groin, which were not of syphilitic origin. The suppuration was in the cellular texture around the glands, which remained indurated three months, assuming a scirrhous character. It was determined to extirpate the tumor. But under fear of the operation the boy left the hospital, returned to the parish of Malacatos, and took a half-pint of a decoction of two ounces of the stick of cundurango in a pint of water, night and morning. At the end of a month he was perfectly well.

In the same year a woman, aged twenty-five, was completely cured of chronic tertiary syphilis with cundurango in sixty days.

A man forty years old was cured of chronic rheumatism and impending paralysis in fifteen days.

Although intelligent and reliable gentlemen of the places visited by Dr. Ayers were very confident of its efficacy, he is of opinion that the diagnosis of some of the most remarkable cases reported is too uncertain, and the use of the remedy has been too limited to establish its curative value.

Three intelligent physicians of considerable experience in the use of cundurango, were of the opinion that, in the treatment of syphilis, it is of less value than the iodide of potassium, and the various preparations of mercury.

Dr. A. Destruge, of Guayaquil, is of opinion that cundurango is very useful in some rheumatisms and secondary syphilitic disorders, but so far as his experience goes of very doubtful influence in cancer.

Mr. Davidson, the house-surgeon of Middlesex Hospital, London, reports that he administered the decoction of cundurango in a case of ulcerated epithelioma of the roof of the mouth; in one of primary cancer of the penis, and secondary infection of the lymphatic glands of both groins: in one of ulcerated epithelioma of the scrotum; and in one of ulcerated scirrhus of the female breast. The cundurango had positively *no effect* upon the progress of the disease in these cases.*

Several cases of cancer, occuring in private practice in Philadelphia, have proved fatal under the use of this article.

Of the many trials of it made by respectable practitioners in different parts of the United States, none is reported successful.

My investigation of the subject leads to a conviction that there is much testimony but no evidence that cundurango, Pseusmagennetus equatoriensis, has a curative influence in cases of cancer. The hope created that it would be found a specific has proved fallacious. The disease is as fatal as ever. The deaths from cancer in Philadelphia have averaged during twelve years 1.362 per cent. of the total mortality, less the still-born. It was lowest in 1865, when it was 0.923 per cent., and highest in 1871, when it was 1.656 per cent. of the total mortality, exclusive of the still-born. During the latter part of this year cundurango was in use.

Besides the cundurangos, I find in the collection forwarded by Dr. Ayers a parcel and samples of the stems and leaves of a small shrub called CHINININGA, which grows on the mountains near Malacatos. The bark of the root of this plant, which is intensely bitter, is said to be a remedy in fevers and skin-diseases of a syphilitic character, in doses of from eight to ten grains three times a day. No flowers were obtained.

At Santa Rosa, Dr. Ayers was told of the recent cure of several cases of phthisis by the use of a fat obtained through the agency of heat from the larvæ of a kind of beetle.

* London Lancet, American edition, February, 1872.

3 c

www.ingramcontent.com/pod-product-compliance
Lightning Source LLC
Chambersburg PA
CBHW022033190326
41519CB00010B/1704